Model Program for Special Inspection

Based on 2006 IBC® Chapter 17

A guideline to assist building officials in the administration and enforcement of special inspection code provisions

International Accreditation Service, Inc.
Leading Accreditors Since 1975

ICC
International Code Council

Model Program for Special Inspection
Based on the 2006 IBC Chapter 17

First Printing: June, 2007
Second Printing: November 2007
Third Printing: January 2008
Fourth Printing: April 2008

ISBN 978-1-58001-569-1

INTERNATIONAL CODE COUNCIL®

The information contained in this document is believed to be accurate; however, it is being provided for informational purposes only and is intended for use only as a guide. Publication of this document by the ICC should not be construed as the ICC engaging in or rendering engineering, legal or other professional services. Use of the information contained in this publication should not be considered by the user as a substitute for the advice of a registered professional engineer, attorney or other professional. If such advice is required, you should seek the services of a registered professional engineer, licensed attorney or other professional.

Printed in the United States of America

FOREWORD

This document provides the building official with guidance on the administration and implementation of the special inspection requirements of Section 1704 of the *2006 International Building Code*® (IBC®). The guidance is based on recommended practices and the consensus of building officials, design professionals, and inspection and testing agencies. Duties and responsibilities of the building official, special inspector, project owner, engineer or architect of record, and contractor are defined in this guide.

In the past, special inspection tasks were often delegated to the project owner, architect or engineer of record, and sometimes building contractors. In the 21st century, the responsibilities of the building official in providing oversight of special inspection activities is driven by several factors, including legal liability, insistence on competent performance and completion of work as specified and approved, and more involvement of local government. Accordingly, the use of certified individuals or accredited organizations is becoming increasingly prevalent and necessary.

Under the IBC, special inspection is not a discretionary activity, nor does it replace required structural observations as required by the design professional in responsible charge. The conditions under which special inspection must be utilized are clearly stated in IBC Section 1704; however, there is a provision for the building official to waive special inspection for minor work.

A special inspector is an individual with specialized skills who observes critical building or structural features identified by the design professional responsible for compliance to plans and specifications approved by the building official. Ensuring the competence of special inspectors is the responsibility of the building official. Although the IBC lacks specific qualification requirements, this fact does not lessen the importance of diligence in the critical process of approving special inspectors.

Recognizing the need to support its member building departments across the U.S., the International Code Council® (ICC®) developed certification and accreditation programs to identify competent individuals and organizations that perform special inspection. ICC Professional Development Services (PDS) certification programs require individuals seeking certification as special inspectors to have demonstrated a level of competence in a manner that will protect the safety and welfare of the public. Individuals may become certified through ICC by participating in a number of seminars and training classes, and passing written examinations. Organizations that employ special inspectors and provide inspection services to the building industry can gain accreditation through International Accreditation Service™ (IAS™), a subsidiary of ICC.

The IAS accreditation program requires special inspection organizations to operate under a documented quality system, to use certified inspectors and to meet comprehensive accreditation requirements, including on-site assessment by subject-matter-experts who are either currently practicing or supervising special inspectors or who are building officials. Successful on-site assessment leads to accreditation for a specific scope of special inspection where the organization has been found competent.

Further information regarding the certification of individuals or accreditation of organizations that perform special inspection can be found on the websites listed below:

IAS	www.iasonline.org
ICC	www.iccsafe.org/certification

ACKNOWLEDGEMENTS

The Model Program for Special Inspection began in 1986 when it was first offered to building officials, jurisdictions, and inspection and testing agencies. It focused on the field inspection aspects of special inspection and is now used as a standard reference for the special inspector exams.

The original program was the culmination of the efforts of building officials, inspectors and inspection/testing agencies associated with California Council of Testing and Inspection Agencies (CCTIA). Since that time, many others have voluntarily given their time and talents to keep this document current and pertinent to the special inspection profession. This document has been reviewed over the years by special inspector exam committees before and after the consolidation of ICC. The following are participants of these committees:

Charles Ballou	Minnesota
Stan Beeler	Ohio
Jeffery Bowers	Washington
James Byrnes	Kansas
Clifford Craig	California
Michelle Craig	California
Fred Deis	Oregon
David Drenth	Minnesota
Roger Evans	Utah
Carl Eriksson	Utah
Fayez Fanik	New Jersey
William Fitzjohn	California
Harry Gleich	South Carolina
Charles Hemsley	California
Jeff Hilfiker	Kansas
Merlyn Isaak	California
Don Johnson	Florida
David Janifer	Virgina
Joe Kane	New York
Roy Keck	Georgia
Ron Lynn	Nevada
Mehrdad Maher	California
Calvin McCall	North Carolina
R.D. Messer	Louisiana
Jim Messersmith	Virginia
Charles Mlodzik	Washington
William Murchinson	North Carolina
Robert Nagin	Florida
Phillip Nishikawa	California
Sam Palmer	Nevada
Gary Pedersen	Washington
Anthony Re	Oregon
Don Riley	Alabama
Gerald Smith	Alabama
Bill Stevens	South Carolina
Thomas Stevens	California
Emile Troup	Maine
Zan Turner	California
W.F. Urbick	Washington
George Voigt	Pennsylvania
Michael Wheeler	California
Kip Williamson	California
David Wismer	Pennsylvania
Daniel Zechmeister	Michigan

Thanks to all who have participated in bringing together a model program for special inspection that can be adapted by others without changing the intent of the performance of the special inspector's work.

The accreditation information that was added in this edition was provided by Pat McCullen, Vice President of IAS.

William Fitzjohn has provided invaluable assistance in reviewing the final version of this document.

Contents

I. SPECIAL INSPECTION – AN OVERVIEW

There are several areas of construction regulated by the *International Building Code*® (IBC®) where special inspection is mandatory. These inspections are to verify that work that is considered critical to life safety and property protection is being or has been constructed according to the approved plans and specifications. These special inspections are in addition to the typical municipal inspections required by the building department that are specified in IBC Section 109 or specific structural observation as may be required by Section 1709. Section 1704 details when special inspection is required and for what types of work. The registered design professional in responsible charge is required to prepare a "Statement of Special Inspections" that outlines the special inspections to be performed.

A special inspector is a person who has been approved by the building official to perform certain types of inspections as detailed in IBC Section 1704. Although Section 1704.1 of the IBC requires the owner or the registered design professional acting as the owner's agent to provide for specially qualified inspectors, the approval of these inspectors is solely the responsibility of the building official.

Special inspection requirements in the 2006 IBC are comprehensive and consist of 16 major categories as follows:

1. **Inspection of fabricators** – Where fabrication of structural load-bearing members and assemblies are being performed on the premises of the fabricator. Note the exception for approved fabricators (See Section 1704.2).

2. **Steel construction** – Steel elements of buildings and structures requiring special inspection are found in Section 1704.3. See Table 1704.3 for detailed information regarding inspections and referenced standards.

   Specific areas listed in the Table 1704.3 for steel construction are as follows:

   - Material verification of high-strength bolts, nuts and washers.
   - Inspection of high-strength bolting.
   - Material verification of structural steel.
   - Material verification of weld filler materials.
   - Inspection of welding for both structural steel and reinforcing steel.
   - Inspection of steel frame joint details for compliance with approved construction documents.

   Note exceptions 1 and 2 listed in Section 1704.3 discussing steel fabrication without heating operations of any kind and continuous and periodic inspection of certain welding operations.

3. **Concrete construction** – The special inspection and verification for concrete construction are found in Section 1704.4. See Table 1704.4 for detailed information regarding inspections.

   Specific areas listed in Section 1704.4 for concrete construction:

   - Reinforcing steel, including prestressing tendons and placement.

- Reinforcing steel welding.
- Bolts to be installed in concrete prior to and during placement of concrete.
- Verifying use of required design mix.
- Sampling fresh concrete and performing slump, air content and fresh concrete temperature at time of making specimens for strength tests.
- Proper application techniques for concrete and shotcrete placement.
- Maintaining specified curing temperature and techniques.
- Prestressed concrete, including application of prestressing forces and grouting of bonded prestressing tendons.
- Erecting precast concrete members.
- Verifying in-situ concrete strength prior to stressing of tendons in post-tensioned concrete and prior to removal of shores and forms from beams and structural slabs.
- Formwork for shape, location and dimensions of concrete members.

Note Exceptions 1, 2, 3, 4 and 5 in Section 1704 discussing footings for buildings three stories or less, nonstructural slabs, foundations and certain exterior concrete features when placed on grade. Be aware of criteria allowing use of these exceptions.

4. **Masonry construction** – The special inspection and verification for masonry construction are found in Section 1704.5. Tables 1704.5.1 and 1704.5.3 show detailed information regarding Level 1 and Level 2 special inspections for masonry construction and whether continuous or period inspection is required.

Specific areas inspected in masonry construction (see Section 1704. 5) are as follows:

- Verifying site-prepared mortar, construction of mortar joints and locations of reinforcement and connectors.
- Verifying size and location of structural elements, type, size and location of anchors, including details of anchorage of masonry to structural members, frames or other construction.
- Verifying specified size, grade and type of reinforcement.
- Verifying welding of reinforcing bars.
- Verifying protection of masonry during cold or hot weather.
- Verifying prior to grouting that grout space is clean and the proportions of site-prepared grout.
- Verifying grout placement is in compliance with code and construction document provisions.
- Inspecting the preparation of any grout specimens, mortar specimens and/or prisms.

- Verifying the compliance with required inspection provisions of the construction documents and the approved submittals.

Note Exceptions 1, 2 and 3 that discuss glass block uses, masonry foundation walls and fireplaces.

5. **Wood construction** – Inspection of the fabrication of wood structural elements and assemblies for high-load diaphragms both prefabricated and field assembled or site built (see Section 1704.6).

   Specific areas of wood construction when high-load diaphragms are used (see Section 1704.6.1):

   - Verification of structural wood panel sheathing for grade and thickness.
   - Verification of nominal size of framing members.
   - Verification of diameter and length of fasteners and spacing of fasteners in the line of structural members and at panel edges.

6. **Soils** – Inspection site for existing conditions, verification of site preparation prior to placement of prepared fill, verification of fill material and maximum lift thicknesses and verification that in-place densities meet the requirements of the approved soils report (see Sections1802.2 and 1704.7).

7. **Pile foundations** – Inspection of installation and testing of pile foundations and record installation of each pile (see Section 1704.8 and Table 1704.8).

8. **Pier foundations** – Inspection installation and testing of pier foundations (see Section 1704.9 and Table 1704.9).

9. **Sprayed fire-resistant materials** – Inspection of fire-resistive material applied to structural elements and decks in accordance with Sections 1704.10.1 through 1704.10.5.

   Specific areas for inspection of sprayed fire-resistant materials are of the following items listed:

   - Verification of structural member surface conditions.
   - Verification of the application of materials per manufacturer's instructions.
   - Verification of the thicknesses and density of applied materials.
   - Verification of the bond strength of applied materials.

10. **Mastic and intumescent fire-resistant coatings** – Inspections for mastic and intumescent for structural elements and decks (see Section 1704.11).

11. **Exterior insulation and finish systems (EIFS)** – See Section 1704.12 for exceptions to inspection when EIFS is applied over water-resistive barriers with a means for draining excess water, and EIFS installed on masonry or concrete.

12. **Special cases** – Inspections, in the opinion of the building official, that are needed because of the use of alternate materials, unusual design or use of materials not having building code approval or needing to meet special manufacturer requirements. See Section 1704.13 for examples where the building official has discretion to require special inspection.

13. **Smoke control** – Inspections involving testing of ductwork during erection, prior to concealment and prior to occupancy for pressure difference, flow measurements and detection and control verification (see Section 1704.14).

14. **Special Inspection for Seismic Resistance** – Section 1705.3 describes special inspection requirements for seismic resistance required in construction systems.

15. **Special Inspection for Wind Requirements** – Section 1705.4 describes special inspection requirements for the main wind force-resisting systems in construction.

II. GENERAL PROGRAM GUIDELINES

A. Purpose of Special Inspection

Special inspection is the monitoring of the materials and workmanship that are critical to the integrity of the building structure. It is the review of the work of the contractors and their employees to ensure that the approved plans and specifications are being followed and that relevant codes and ordinances are being observed. The special inspection process is in addition to those conducted by the municipal building inspector and by the engineer or architect of record as part of periodic structural observation. Special inspectors furnish continuous or periodic inspection as prescribed in IBC Table 1704.3 for that construction which requires their presence (IBC Sections 109.3.9 and 1704).

Good communication between the special inspector and the designers, contractor and building department is essential to project quality assurance. The following sections, B through F, are a combination of building code requirements and the results of years of special inspection work that have resulted in the successful completion of buildings.

B. Duties and Responsibilities of the Special Inspector

The building code requires that a "Statement of Special Inspection" be submitted with the application of the permit. The special inspector should know and understand the scope of the statement prior to beginning special inspections (see the duties of the design professional in responsible charge).

Though not required by code, special inspectors and/or inspection agencies can document acceptance of their responsibilities and scope of work for a project by signing an agreement that includes a detailed schedule of services, such as the Special Inspection and Testing Agreement and the Special Inspection and Testing Schedule included in Appendix A.

Duties of special inspectors and/or inspection agencies include the following:

1. **General requirements.** Special inspectors shall review approved plans and specifications for special inspection requirements. Special inspectors will comply with the special inspection requirements of the enforcing jurisdiction found in the Statement of Special Inspections including work and materials.

2. **Signify presence at jobsite.** Special inspectors shall notify contractor personnel of their presence and responsibilities at the jobsite. If required by the building official, they shall sign in on the appropriate form posted with the building permit. An example of a special inspection jobsite record is included in Appendix A.

3. **Observe assigned work.** Special inspectors shall inspect all work according to the Statement of Special Inspections for which they are responsible for conformance with the building department approved (stamped) plans and specifications and applicable provisions of IBC Section 1704.

4. **Report nonconforming items (discrepancies).** Special inspectors shall bring all nonconforming items to the immediate attention of the contractor. If any such item is not resolved in a timely manner or is soon to be incorporated into the work, the engineer or architect of record and the building official should be notified immediately and the item noted in the special inspector's written report (Section 1704.1.2). Some jurisdictions may require this report to be a separate, individual report from the progress reports. The building official may require this report to be

posted in a conspicuous place on the job site. The special inspector should include in the report, as a minimum, the following information about each nonconforming item:

- Description and exact location.
- Reference to applicable detail of approved plans/specifications.
- Name and title of each individual notified and method of notification.
- Resolution or corrective action taken.

5. **Provide timely progress reports.** The special inspector shall complete written inspection reports for each inspection visit and provide the reports on a timely basis as determined by the building official. The special inspector or inspection agency shall furnish these reports directly to the building official and to the design professional in charge (Section 1704.1.2). These reports should be organized on a daily format and may be submitted weekly at the option of the building official. In these reports, special inspectors should:

   - Describe inspections and tests made with applicable locations and whether the work meets the requirements of the statement of special inspections.
   - Indicate nonconforming items (discrepancies) and how they were resolved.
   - List unresolved items, parties notified, and time and method of notification.
   - Itemize changes authorized by engineer or architect of record if not included in nonconforming items.

   Examples of report forms are included in Appendix A.

6. **Submit final report.** Special inspectors or inspection agencies shall submit a final signed report to the building department stating that all items requiring special inspection and testing by the Statement of Special Inspection were fulfilled and reported and, to the best of their knowledge in conformance with the approved plans and specifications (Section 1704.1.2). Jurisdictions may also require the design professional in responsible charge to sign the report before it is submitted to the building official. Items not in conformance, unresolved items or any discrepancies in inspection coverage (i.e., missed inspections, periodic inspection when continuous was required, etc.) should be specifically itemized in this report. An example of a final report form is included in Appendix A.

C. Duties and Responsibilities of the Project Owner

The project owner, the engineer or architect of record, or an agent of the owner is responsible for funding special inspection services. Measures should be taken to ensure that the scope of work and duties of the special inspector as outlined in the Statement of Special Inspection are not compromised.

D. Duties and Responsibilities of the Design Professional in Responsible Charge

The design professional in responsible charge should be a consenting party by written acknowledgment of special inspection and testing agreements. An example of such an agreement is included in Appendix A.

The design professional in responsible shall prepare a "Statement of Special Inspections" to be submitted with the permit application (Sections 106.3.4 and 1704.1.1). The content of the statement is to contain the following information (Section 1705.2).

1. Materials, systems, components and work required to have special inspection. Duties of the building official and the registered design professional responsible for each part of the work are to be stated.

2. Type and extent of each special inspection and the name of individual or firms performing the inspections.

3. Type and extent of each test.

4. Special inspection of required seismic resistance systems and components (Section 1705.3).

5. Special inspection of required wind resistance systems and components (Section 1705.4).

6. Requirements for structural observation. See Sections 106.3.4.1 and 1709 for information regarding specific requirements.

The following are elements of good practice and could be included in the statement of special inspections.

7. Provide name and contact information of special inspector or special inspection agency. Subject to the approval of the building official, special inspectors holding current certification by ICC in the discipline in which they will be inspecting can be considered qualified. Likewise, special inspection agencies holding current accreditation with IAS (See Appendix IV) with the appropriate scope of accreditation for the disciplines to be inspected can be considered qualified. The choice of special inspectors or special inspection agencies should include the following considerations:

    - Project size and complexity – experience with similar projects.
    - Inspection staffing – sufficient qualified inspectors.
    - Site location – proximity of inspection and testing facilities.
    - Offsite inspection – capabilities for inspection at remote locations.

8. Completion of a special inspection and testing agreement and schedule as shown in the Appendix A examples is a simple method of fulfilling the requirement for preparation of a special inspection program that can be easily reviewed by the building official.

9. Respond to field discrepancies. The engineer or architect of record shall respond to special inspector reports of uncorrected noncomplying (discrepancies) items and shall approve remedial measures.

10. Review shop drawings and submit revisions to approved plans. The design professional in responsible charge shall acknowledge and approve shop drawings that may detail structural information. The design professional shall submit to the building official and the special inspector/ inspection agency written approval of any verbally approved deviations from the approved plans and shall submit revised plans for building official approval (Section 106.3.4). It is becoming more common for the building official to require the signature of the design professional on all final documents.

It is good practice that the design professional in responsible charge be a signatory of the statement of special inspection and of any subsequent agreements affecting the work of special inspection of the project.

E. Duties and Responsibilities of the Contractor

Each contractor responsible for construction of the structural systems described in Section 1706 shall submit a written statement of responsibility to the building official and the owner prior to commencement of the work on the system or component. The contractor's statement of responsibility is to contain the following:

1. Acknowledgement of awareness of the special requirements contained in the statement of special inspections;

2. Acknowledgement that control will be exercised to obtain conformance with the construction documents approved by the building official;

3. Procedures for exercising control within the contractor's organization, the method and frequency of reporting and the distribution of the reports; and

4. Identification and qualifications of the person(s) exercising such control and their position(s) in the organization.

In addition to the responsibilities mentioned by code, the contractor should:

5. Notify the special inspector. Adequate notice shall be provided so that the special inspector has time to become familiar with the project.

6. Provide access to approved plans. The contractor is responsible for providing the special inspector with access to approved plans (IBC Section 106.3.1).

7. Retain special inspection records. When required by the building official, the contractor is responsible for retaining at the job-site all special inspection records submitted by the special inspector and providing these records for review by the building department's inspector upon request.

F. Duties and Responsibilities of the Building Official

Of all the team members involved in the construction process, the building official is the only one with the legal authority to enforce the special inspection provisions of the code (Section 104.4). The employment of a special inspector or agency shall not relieve the building department of responsibility for special inspections or structural observations as required by the code and stated in the Statement of Special Inspections. Building department inspections of items also requiring special inspection should not be signed off without the concurrence of the special inspector.

The specific duties and responsibilities of the building official relating to special inspection include the following:

1. Review submittal documents for compliance with special inspection requirements. The building official is charged with the legal authority to review the plans, specifications pertaining to work and materials, special inspection program and other submittal documents for compliance with code requirements (Sections 104.2 and 106.3 through 106.5). Approval for materials should in accordance with Sections 104.9 and 104.11.

2. Approve Fabricator(s) used for building components installed on-site.

3. Approve special inspection program. The building official is responsible for approving the special inspection program submitted by the design professional in responsible charge (Section 106.1) and may require a preconstruction conference to review the program with all applicable members of the construction team.

4. Approve special inspectors/inspection agencies. The building official is responsible for determining competence of special inspectors for the types of work they will be inspecting (Section 1704). Suggested qualifications for special inspectors are contained in Part III of this guide.

5. Monitor special inspection activities. The building official should monitor the special inspection activities at the job-site to assure that qualified special inspectors are performing their duties when work requiring special inspection is in progress (Section 1704.1.2).

6. Issuance of stop work orders. The building official is recognized as having the authority to stop work at the job-site (Section 114), however, the special inspector does not have this authority. Stop work orders most likely will be issued in response to a discrepancy report that has been written.

7. Approval to proceed. There are certain points of completion where work shall not proceed until approval by the building official has been given.

8. Review inspection reports. The building official receives and reviews special inspection progress reports and final reports for conformance with the approved plans, specifications and workmanship provisions of the code (Section 1704.1.2).

9. The building official should not perform the final inspection and issue certificate of occupancy for a project (IBC Section 109.3.10) until the final special inspection report has been reviewed and approved.

(This page is intentionally left blank.)

III. RECOMMENDED SPECIAL INSPECTOR QUALIFICATIONS

A. Code Knowledge and Plan Reading Competency Standards

The building official has the responsibility to ascertain that special inspectors and/or agencies are qualified for the type(s) of inspection required. In some geographic areas, cooperative jurisdictional programs for the recognition of competent inspectors/agencies are available to assist in satisfying this requirement. Qualification standards may include:

1. Individuals maintaining current certification by ICC as a special inspector for the discipline(s) for which he or she is requesting approval.

2. Agencies maintaining current accreditation as a special inspection agency by IAS with a scope of accreditation covering the discipline(s) for which the agency is requesting approval.

3. Written or oral examination to verify the applicant's knowledge of jurisdictional procedures and code requirements.

B. Experience Standards

The specialized knowledge and skills needed to perform special inspection should be obtained by experience in some aspect of the discipline being inspected. Industry standards typically require 1–5 year(s) of verifiable practical experience, depending upon the discipline and the level of inspection, a portion of which may be satisfied by applicable technical education. There are also provisions for trainees who may work under the direct supervision of qualified special inspectors. An example of detailed guidelines for special inspector experience standards is contained in Appendix C.

The building official may require that special inspector applicants provide evidence of experience through the following and/or other means:

1. Written references verifying related work experience; and/or

2. Personal interview to evaluate the applicant's work experience and suitability to be a special inspector.

3. Evidence of compliance with a recognized industry experience guideline (see example in Appendix C).

C. Performance Standards

The building official may use the model code organization job task lists for special inspector certification exams as guidelines for evaluating the performance of special inspectors in each of the applicable disciplines. These lists are included in Appendix B and contain comprehensive tasks that may be performed by special inspectors on large and complex projects. Actual tasks required on specific projects will typically be fewer.

(This page is intentionally left blank.)

IV. RECOMMENDED TESTING AND INSPECTION AGENCIES ACCREDITATION

A. Special Inspection Agencies Program

In response to a request from several ICC governmental members, International Accreditation Service (IAS) has developed an accreditation program for agencies providing the special inspections that are required for specific construction projects under Chapter 17 of the 2006 IBC. The IAS Special Inspection Agency Accreditation Program is based primarily on the requirements of the IBC and the applicable portions of ISO/IEC 17020, General Requirements for the Operation of Various Bodies Performing Inspection. The program requires special inspection agencies to operate under a quality management system that is documented in a manual, and also requires the agency to be assessed in the field to determine if it is competent to perform specific inspections or types of inspections. IAS accreditation is based on the assessment of a special inspection agency's inspection procedures, the competence of its inspection staff and its reporting procedures.

B. Testing Laboratories Program

Industry depends on reliable assessment of testing and calibration laboratories, inspection agencies and fabricator inspection programs. Accurate test, calibration and inspection results are vital for the protection of public health and safety and to facilitate trade. IAS operates nationally and internationally-recognized accreditation programs, some of which are specifically tailored to the needs of the building official. The IAS laboratory accreditation program is modeled on international standard ISO/IEC Standard 17025 and is recognized globally by the International Laboratory Accreditation Cooperation (ILAC), the Asia Pacific Laboratory Accreditation Cooperation (APLAC) and nationally by the National Cooperation for Laboratory Accreditation (NACLA). Laboratory accreditation, traceability of measurements and uniform testing requirements minimize technical barriers to trade and reduce the financial burden on manufacturers. Evaluation of a product, material or piece of equipment should be based on reports from accredited testing laboratories that are supported by accredited calibration laboratories to ensure traceable and repeatable measurements.

C. Public, Affected Parties and Regulator Access

The IAS accreditation programs are monitored by an accreditation committee that is composed of code regulators who approve all accreditation criteria. Through its accreditation committee, IAS follows a public hearing process that allows for input from regulators, laboratories, inspection agencies and other interested parties. Members of the accreditation committee are appointed by the IAS Board of Directors. The accreditation committee provides a transparent process that provides access to all interested parties. On technical matters IAS maintains two formally appointed Technical Advisory Councils (TACs) consisting of experts drawn from accredited calibration and testing laboratories and special inspection agencies. Each TAC provides technical guidance to IAS on matters relating to calibration and testing or special inspection as the need arises.

(This page is intentionally left blank.)

APPENDIX A - SECTION 1

Examples of Special Inspection Forms, Schedules and Agreements

Municipal Agency Forms

- Special Inspection and Testing Agreement
- Special Inspection and Testing Schedule
- Special Inspection Record

(THESE FORMS MAY BE PHOTOCOPIED)

(This page is intentionally left blank.)

STATEMENT OF SPECIAL INSPECTIONS AGREEMENT

To permit applicants of projects requiring special inspection and/or testing per Section 1704 of the *International Building Code* (IBC):

Project Address: ____ Permit No.: _______________

BEFORE A PERMIT CAN BE ISSUED, two (2) copies of this agreement, including the Statement of Special Inspection and the Special Inspection and Testing Schedule with the required acknowledgments shall be completed by the owner, or registered design professional in responsible charge acting as the owner's agent. A preconstruction conference with the parties involved may be required to review the special inspection requirements and procedures.

APPROVAL OF SPECIAL INSPECTORS: Special inspectors may have no financial interest in projects for which they provide special inspection. Special inspectors shall be approved by the building department prior to performing any duties. Special inspectors shall submit their qualifications and are subject to personal interviews for prequalification. Special inspectors shall display approved identification, as stipulated by the building official, when performing the function of special inspector.

Special inspection and testing shall meet the minimum requirements of Section 1704 through 1707 of the *International Building Code.* The following conditions are also applicable:

A. Duties and Responsibilities of the Special Inspector

1. **General requirements.** Special inspectors shall review approved plans and specifications for special inspection requirements. Special inspectors will comply with the special inspection requirements of the enforcing jurisdiction found in the Statement of Special Inspections including work and materials.

2. **Signify presence at job-site.** Special inspectors shall notify contractor personnel of their presence and responsibilities at the job-site. If required by the building official, they shall sign in on the appropriate form posted with the building permit.

3. **Observe assigned work.** Special inspectors shall inspect all work according to the Statement of Special Inspections for which they are responsible for conformance with the building department approved (stamped) plans and specifications and applicable provisions of IBC Section 1704.

4. **Report nonconforming items (discrepancies).** Special inspectors shall bring all nonconforming items to the immediate attention of the contractor. If any such item is not resolved in a timely manner or is about to be incorporated into the work, the engineer or architect of record and the building official should be notified immediately and the item noted in the special inspector's written report (Section 1704.1.2). The building official may require this report to be posted in a conspicuous place on the job-site. The special inspector should include in the report, as a minimum, the following information about each nonconforming item:

   - Description and exact location.
   - Reference to applicable detail of approved plans/specifications.
   - Name and title of each individual notified and method of notification.
   - Resolution or corrective action taken.

5. **Provide timely progress reports.** The special inspector shall complete written inspection reports for each inspection visit and provide the reports on a timely basis as determined by the building official. The special inspector or inspection agency shall furnish these reports directly to the building official and to the

design professional in charge (Section 1704.1.2). These reports should be organized on a daily format and may be submitted weekly at the option of the building official. In these reports, special inspectors should:

- Describe inspections and tests made with applicable locations.
- Indicate nonconforming items (discrepancies) and how they were resolved.
- List unresolved items, parties notified, and time and method of notification.
- Itemize changes authorized by engineer or architect of record if not included in nonconforming items.

6. **Submit final report.** Special inspectors or inspection agencies shall submit a final signed report to the building department stating that all items requiring special inspection and testing by the Statement of Special Inspection were fulfilled and reported and, to the best of their knowledge in conformance with the approved plans and specifications (Section 1704.1.2). Some jurisdictions also require the design professional in responsible charge to sign the report before it is submitted to the building official. Items not in conformance, unresolved items or any discrepancies in inspection coverage (i.e., missed inspections, periodic inspection when continuous was required, etc.) should be specifically itemized in this report.

B. Owner Responsibilities. The project owner, the engineer or architect of record, or an agent of the owner is responsible for funding special inspection services. Measures should be taken to ensure that the scope of work and duties of the special inspector as outlined in the Statement of Special Inspection are not compromised.

C. Registered Design Professional in Responsible Charge

The design professional in responsible charge should be a consenting party by written acknowledgment of special inspection and testing agreements. The design professional in responsible charge has many duties and responsibilities related to special inspection, including the following:

The Statement of Special Inspections has been submitted with the permit application. The content of the statement contains the following information.

1. Materials, systems, components and work required to have special inspection. Duties of the building official and the registered design professional responsible for each part of the work are to be stated.

2. Type and extent of each special inspection and the name of individual or firms performing the inspections.

3. Type and extent of each test.

4. Special inspection of required seismic resistance systems and components.

5. Special inspection of required wind resistance systems and components.

6. Provide name and contact information of special inspector or special inspection agency.
   Subject to the approval of the building official, special inspectors holding current certification by ICC in the discipline in which they will be inspecting can be considered qualified within the appropriate scope of accreditation for the disciplines to be inspected can be considered qualified. The choice of special inspectors or special inspection agencies should include the following considerations:

   - Project size and complexity – experience with similar projects.
   - Inspection staffing – sufficient qualified inspectors.
   - Site location – proximity of inspection and testing facilities.
   - Off-site inspection – capabilities for inspection at remote locations.

7. Completion of a special inspection and testing agreement and schedule as shown in the Appendix A examples is a simple method of fulfilling the requirement for preparation of a special inspection program that can be easily reviewed by the building official.

8. Respond to field discrepancies. The engineer or architect of record shall respond to special inspector reports of uncorrected non-complying (discrepancies) items and shall approve remedial measures.

9. Review shop drawings and submit revisions to approved plans. The design professional in responsible charge shall acknowledge and approve shop drawings that may detail structural information. The design professional shall submit to the building official and the special inspector/inspection agency written approval of any verbally approved deviations from the approved plans and shall submit revised plans for building official approval (Section 106.3.4).

D. Contractor Responsibilities

Submit a written statement of responsibility to the building official and the owner prior to commencement of the work on the system or component. The contractor's statement of responsibility is to contain the following:

1. Acknowledgement of awareness of the special requirements contained in the statement of special inspections;

2. Acknowledgement that control will be exercised to obtain conformance with the construction documents approved by the building official;

3. Procedures for exercising control within the contractor's organization, the method and frequency of reporting and the distribution of the reports; and

4. Identification and qualifications of the person(s) exercising such control and their position(s) in the organization.

Suggested job-site protocol for special inspection:

1. Notify the special inspector. Adequate notice shall be provided so that the special inspector has time to become familiar with the project.

2. Provide access to approved plans. The contractor is responsible for providing the special inspector with access to approved plans.

3. Retain special inspection records. When required by the building official, the contractor is responsible for retaining at the job-site all special inspection records submitted by the special inspector and providing these records for review by the building department's inspector upon request.

E. Building Department Responsibilities

1. Review submittal documents for compliance with special inspection requirements as outlined in the Statement of Inspection. The building official is charged with the legal authority to review the plans, specifications, special inspection program and other submittal documents for compliance with code requirements.

2. Approve fabricator(s) used for building components installed on-site.

3. Approve special inspection program. The building official is responsible for approving the special inspection program submitted by the design professional in responsible charge (Section 106.1) and may require a preconstruction conference to review the program with all applicable members of the construction team.

4. Monitor special inspection activities. The building official should monitor the special inspection activities at the jobsite to assure that qualified special inspectors are performing their duties when work requiring special inspection is in progress.

5. Monitor special inspection activities. The building official should monitor the special inspection activities at the jobsite to ensure that qualified special inspectors are performing their duties when work requiring special inspection is in progress.

6. Issuance of stop work orders. The building official is recognized as having the authority to stop work at the job-site.

7. Approval to proceed. There are certain points of completion where work shall not proceed until approval by the building official has been given.

8. Review inspection reports. The building official receives and reviews special inspection progress reports and final reports for conformance with the approved plans, specifications and workmanship provisions of the code.

9. Perform final inspection. The building official should not perform the final inspection and approval for a project until the final special inspection report has been reviewed and approved.

ACKNOWLEDGMENTS

I have read and agree to comply with the terms and conditions of this agreement.

Owner:

________________________________By: ________________________Date: ___________

Registered Design Professional in Responsible Charge:

________________________________By: ________________________Date: ___________

Contractor:

________________________________By: ________________________Date: ___________

Special Inspector / Special Inspection Agency*:

________________________________By: ________________________Date: ___________

Others as required by Building Official:

________________________________By: ________________________Date: ___________

ACCEPTED FOR THE BUILDING DEPARTMENT

By: ____________________________ Date: ___________

* This signature may be that of the Responsible Professional Engineer within the special inspection agency.

STATEMENT OF SPECIAL INSPECTIONS SCHEDULE

(This schedule should match the Statement of Special Inspections)

Where the option between continuous and periodic inspection is possible, circle the option required.

Steel Construction – Verification / Inspection See Table 1704.3 – Required Verification and Inspection of Steel Construction	**Continuous**	**Periodic**
1. High-strength bolts, nuts and washers		
a. [ ] Markings	---	X
b. [ ] Manufacturer's certificate of compliance	---	X
2. High-strength bolting		
a. [ ] Bearing-type connections	---	X
b. [ ] Slip-critical connections	X	X
3. Structural Steel		
a. [ ] Identification Markings	---	---
b. [ ] Manufacturer's certified mill reports	---	---
4. Weld filler		
a. [ ] Identification Markings	---	---
b. [ ] Manufacturer's certificate of compliance	---	---
5. Welding –		
a. **Structural Steel**		
1. [ ] Complete and partial penetration groove welds	X	---
2. [ ] Multi-pass fillet welds	X	---
3. [ ] Single-pass fillet welds > 5/16"	X	---
4. [ ] Single-pass fillet welds ≤ 5/16"	---	X
5. [ ] Floor and roof deck welds	---	X
b. **Reinforcing Steel**		
1. [ ] Weldability of reinforcing steel other than ASTM A706	---	X
2. [ ] Reinforcing steel – resisting flexural and axial forces in intermediate and special moment frames, and boundary elements of special reinforced concrete shear walls and shear reinforcement.	X	---
3. [ ] Shear reinforcement	X	---
4. [ ] Other reinforcing steel	---	X
6. Steel Frame Joint Details		X
a. [ ] Details – bracing, stiffening, etc.	---	---
b. [ ] Member Locations	---	---
c. [ ] Application of joint details at each connection	---	---
7. Other		

Concrete Construction – Verification / Inspection See Table 1704.4 – Required Verification and Inspection of Concrete Construction		
1. [ ] Reinforcing steel, including prestressing tendons, and placement	---	X
2. [ ] Reinforcing steel welding	---	---
3. [ ] Bolts installed in concrete	X	---
4. [ ] Required mix design	---	X
5. [ ] Fresh concrete sampling / testing	X	---
6. [ ] Placement of concrete and shotcrete	X	---
7. [ ] Curing – temperature and techniques	---	X
8. Prestressed concrete		
a. [ ] Application of prestressing forces	X	---
b. [ ] Grouting of bonded tendons in seismic-force-resisting system.	X	
9. [ ] Erection of precast members	---	X
10. [ ] Verification of in-situ concrete strengths	---	X
11. [ ] Formwork shape, location, and dimensions	---	X
12. Other		

Masonry Level 1 – Verification / Inspection See Table 1704.5.1 – Level 1 Special Inspection	**Continuous during task**	**Periodic during task**
1. As construction begins ensure compliance of:		
a. [ ] Proportions of site-prepared mortar.	---	X
b. [ ] Construction of mortar joints.	---	X
c. [ ] Location of reinforcement, connectors,	---	X
d. [ ] Prestressing technique	---	X
e. [ ] Grade and size of prestressing tendons and anchorages	---	X
2. Inspection shall verify:		
a. [ ] Size and location of structural elements.	---	X
b. [ ] Type, size and location of anchors, including other details of anchorage of masonry to structural members, frames or other construction.	---	X
c. [ ] Specified size, grade and type of reinforcement.	---	X
d. [ ] Welding of reinforcing bars.	X	---
e. [ ] Protection of masonry during cold weather (temperature below 40° F) or hot weather (temperature above 90° F).	---	X
f. [ ] Application and measurement of pretressing force.	---	X
3. Prior to grouting, verify following to ensure compliance:		
a. [ ] Grout space is clean.	---	X
b. [ ] Placement of reinforcement and connectors, and, prestressing tendons and anchorages.	---	X
c. [ ] Proportions of site-prepared grout and prestressing grout for bonded tendons.	---	X
d. [ ] Construction of mortar joints.	---	X
4. Grout placement verified to ensure compliance with code and construction document provisions.	X	---
a. [ ] Grouting of prestressing bonded tendons.	X	---
5. [ ] Preparation of any required grout specimens, mortar specimens and/or prisms shall be observed.	X	---
6. [ ] Compliance with required inspection provisions of the construction documents and the approved submittals shall be verified.	---	X
7. Other		

Masonry Level 2 – Verification / Inspection See Table 1704.5.3 – Level 2 Special Inspection	**Continuous during task**	**Periodic during task**
1. From the beginning of masonry construction, the following shall be verified to ensure compliance:		
a. [ ] Proportions of site prepared mortar, grout and prestressing grout for bonded tendons.	---	X
b. [ ] Placement of masonry units and construction of mortar joints.	---	X
c. [ ] Placement of reinforcement, connectors and prestressing tendons and anchorages.	---	X
d. [ ] Grout space prior to grouting	X	---
e. [ ] Placement of grout.	X	---
f. [ ] Placement of prestressing grout.	X	---
2. The inspection program shall verify:		
a. [ ] Size and location of structural elements.	---	X
b. [ ] Type, size and location of anchors, including other details of anchorage of masonry to structural members, frames or other construction.	X	---
c. [ ] Specified size, grade and type of reinforcement.	---	X
d. [ ] Welding of reinforcing bars.	X	---
e. [ ] Protection of masonry during cold weather (temperature below 40° F) or hot weather (temperature above 90° F).	---	X
f. [ ] Application and measurement of pretressing force.	X	---

3. [ ] Preparation of any required grout specimens, mortar specimens and/or prisms shall be observed.	X	---
4. [ ] Compliance with required inspection provisions of the construction documents and the approved submittals shall be verified.	---	X
5. Other		

Soils – Verification / Inspection See Table 1704.7	**Continuous during task**	**Periodic during task**
1. [ ] Verify materials below footings are adequate to achieve the design bearing capacity.	---	X
2. [ ] Verify excavations are extended to proper depth and have reached proper material.	---	X
3. [ ] Verify classification and testing of controlled fill have been performed.	---	X
4. [ ] Verify use of proper materials, densities and lift thicknesses during placement and compaction of controlled fill.	X	---
5. [ ] Prior to placement of controlled fill, observe subgrade and verify that site has been prepared properly.	---	X
6. Other		

Pile Foundations – Verification / Inspection See Table 1704.8	**Continuous during task**	**Periodic during task**
1. Verify pile materials, sizes and lengths comply with requirements.	X	---
2. Determine capacities of test piles, and conduct additional load tests, as required.	X	---
3. Observe driving operations and maintain complete and accurate records for each pile.	X	---
4. Verify placement locations and plumbness, confirm type and size of hammer, record number of blows per foot of penetration, and required penetrations to achieve design capacity. Record tip and butt elevations and document pile damage. Required penetration is usually determined by the registered design professional for this work.	X	---
5. For steel piles, verify additional inspections have been made in accordance with Section 1704.3 and Table 1704.3.	---	---
6. For concrete piles and concrete-filled piles, verify that additional inspections have been made in according with Section 1704.4 and Table 1704.4.	---	---
7. For specialty piles, perform additional inspections as determined by the registered design professional in responsible charge.	---	---
8. For augered uncased piles and caisson piles, perform inspections in accordance with Section 1704.9.	---	---
9. Other		

Pier Foundations – Verification / Inspection See Table 1704.9		
1. Observe drilling operations and maintain complete and accurate records for each pier.	X	---
2. Verify placement locations and plumbness, confirm pier diameters, bell diameters (if applicable), lengths, embedment into bedrock (if applicable) and adequate end bearing strata capacity.	X	---
3. For concrete piles and concrete-filled piles, verify that additional inspections have been made in according with Section 1704.4 and Table 1704.4.	---	---
4. For masonry piles, verify that additional inspections have been made in according with Section 1704.5 and Tables 1704.5.1 and 1704.5.2.	---	---

Other materials – Verification / Inspection		

Other Work – Verification / Inspection		

Note – As schedules are developed, space for the special inspector, date of inspection, and the design profession in responsible charge should be provided as well as others who may need to recognized the inspections have been complete.

SPECIAL INSPECTION RECORD

City of ______________________________

Inspection Agency: ________________________
Project Address: ________________________

Building Permit No.: ________________________

NOTE: Each special inspector is to record their presence on the job-site for each day's inspection. Post this card adjacent to building permit inspection report card. Weekly reports are to be submitted by each special inspection/inspection agency to the building department.

When attached to the job inspection record card, this card becomes a part of the inspection record.

INSPECTION TYPE	SPECIAL INSPECTOR	ID NO.	DATE	NOTES	TIME	
					ARR	LEFT

(This page is intentionally left blank.)

APPENDIX A – SECTION 2

Examples of Special Inspection Forms, Schedules and Agreements

Special Inspection Forms

- Daily Report Form
- Weekly Report Form
- Discrepancy Notice
- Final Report Form

(THESE FORMS MAY BE PHOTOCOPIED)

(This page is intentionally left blank.)

SPECIAL INSPECTION DAILY REPORT

City/County of ______________________________ Permit No.:____________ Date :________________

Project Name/Address: ____

Inspection Type(s)/Coverage: ____

____

☐ Continuous ☐ Periodic; frequency: ____________________________

Inspections made, including locations: ____

____

____

____

____

____

____

____

____

Tests performed: ____

____

____

____

____

Items requiring 1) correction, 2) correction of previously listed items, and 3) previously listed uncorrected items: ____

____

____

____

Changes to approved plans authorized by engineer or architect of record: ________________________

____

____

____

____

Comments: ____

____

____

____

____

To the best of my knowledge, work inspected was in accordance with the building department approved plans, specifications and applicable workmanship provisions of the IBC except as noted above.

Signed: ______________________________ Inspection Agency*: __________________________

Print full name: __________________________ ID / Certificate Number: ______________________

*** Building official may require the signature / stamp of agency engineer responsible for special inspection.**

(This page is intentionally left blank.)

SPECIAL INSPECTION WEEKLY REPORT

City/County of ______________________________ Permit No.:____________ Date :________________

Project Name/Address: ____

Inspection Type(s)/Coverage: ____

____

☐ Continuous ☐ Periodic; frequency: ____________________

Total inspection time each day:

Dates							
Hours							
Inspector							

Inspections made, including locations:____

____

____

____

____

____

Tests performed: ____

____

____

____

____

Items requiring 1) correction, 2) correction of previously listed items, and 3) previously listed uncorrected items: ____

____

____

____

Changes to approved plans authorized by engineer or architect of record: ____________________

____

____

____

Comments: ____

____

____

____

To the best of my knowledge, work inspected was in accordance with the building department approved plans, specifications and applicable workmanship provisions of the IBC except as noted above.

Signed: ______________________________ Inspection Agency*: ______________________

Print full name: ______________________________ ID / Certificate Number: ____________________

cc: Building Department

Engineer/Architect

*** Building official may require the signature / stamp of agency engineer responsible for special inspection.**

(This page is intentionally left blank.)

SPECIAL INSPECTION DISCREPANCY NOTICE

City/County of ______________________________ Permit No.:____________ Date :________________

Project Name/Address: ____

Inspection Type(s)/Coverage: ____

____

☐ Continuous ☐ Periodic; frequency: ____________________________

Notice delivered to: ☐ Contractor ☐ Engineer/Architect ☐ Building Department

The following discrepancies require correction and inspection approval prior to proceeding with this phase of the work:

____

____

____

____

____

____

____

____

____

____

____

____

____

____

____

____

____

____

____

____

____

____

____

____

Signed: ________________________________ Inspection Agency*: _________________________

Print full name: __________________________ ID / Certificate Number:_______________________

*** Building official may require the signature / stamp of agency engineer responsible for special inspection.**

DO NOT REMOVE THIS NOTICE

Post with building permit inspection record card

(This page is intentionally left blank.)

SPECIAL INSPECTION FINAL REPORT

City/County of ______________________________ Permit No.:____________ Date: ________________

Attention: __________________________________

Project Name/Address: ____

____

In accordance with Section 1704 of the *International Building Code*, special inspection has been provided for the following items:

____

____

____

____

____

____

____

____

____

____

____

____

____

____

Based upon inspections performed and our (my) substantiating reports, it is our (my) professional judgment that, to the best of our (my) knowledge, the inspected work was performed in accordance with the approved plans, specifications and applicable workmanship provisions of the *International Building Code*.

Signed: _________________________________ Inspection Agency*: _________________________

Print full name: ____________________________ ID / Certificate Number:______________________

*** Building official may require the signature / stamp of agency engineer responsible for special inspection.**

cc: Client/Project Owner
    Engineer/Architect

(This page is intentionally left blank.)

APPENDIX B

Job Task Lists for Special Inspectors:

- Reinforced Concrete
- Prestressed Concrete
- Structural Masonry
- Structural Steel and Bolting
- Structural Steel and Welding
- Spray-applied Fireproofing

(This page is intentionally left blank.)

JOB TASK LISTS FOR SPECIAL INSPECTORS

These task lists are the basis of the ICC certification exams. Information for the exams are found in the 2007 ICC National Certification Examination Program Bulletin found at: http://www.iccsafe.org/certification/2007/pdf/2007Bulletin_CodeProfessionals.pdf

The job task lists are developed by the exam development committee specifically assigned to each exam. The tasks describe the work performed by special inspectors as it is practiced in the field. Other Tasks related to this work may be helpful or extend the professional ability of the inspector, however, these tasks are the minimum required to protect the public health and safety. Minimum requirements for certification exams are mandated by the Equal Employment Opportunity Commission (EEOC), and, accepted professional certification standards as developed by the American Psychological Association (APA), Council on Licensure, Enforcement, and Regulation (CLEAR), and National Organization of Competency Assurance (NOCA).

The building official and others may desire to review these tasks in reviewing the proposed special inspection activities for a particular project and evaluating the qualifications of special inspectors. Also, the minimum qualifications listed in the IAS Document AC291 - Accreditation Criteria for IBC Special Inspection Agencies outlined in Appendix C examples of qualifications.

A. Reinforced Concrete Special Inspector

1. General Requirements

Review approved plans and specifications for special inspection requirements. Comply with special inspection requirements of the enforcing jurisdiction. Notify the contractor of deviations from approved plans and specifications. If the deviations are uncorrected, notify the architect or engineer of record, and the building official. Submit progress reports to the architect or engineer of record and the building official, describing tests that were performed and compliance of work. Submit final summary report stating whether work requiring special inspection was in conformance with the approved plans and applicable provisions of the building code.

2. Concrete Quality

Verify that individual batch tickets indicate delivery of the approved mix as specified. Verify time limits of mixing, total water added, and proper consistency and workability for placement. Determine the required type, quantity and frequency of tests to be performed on fresh and hardened concrete. Observe sampling of concrete, field testing of fresh concrete and making of test specimens. Provide or arrange for proper specimen identification, site storage and protection, and transportation to the testing laboratory. Provide or arrange for communication of field-testing results to the architect or engineer of record and to the building official.

3. Reinforcement

Verify that reinforcing steels are of the type, grade and size specified and are in conformance with acceptable quality standards. Ensure that reinforcing steel is free of oil, dirt and rust and that steel is properly coated and/or sheathed as specified. Verify that reinforcing steels are located within acceptable tolerances and are adequately supported and secured to prevent displacement during concrete placement. Verify that minimum concrete cover is provided. Verify that placement of reinforcing steel (or ducts) complies with required spacing, profile and quantity requirements, as indicated by both the approved plans and installation drawings. Verify that hooks, bends, ties, stirrups and supplemental reinforcement are fabricated and placed as specified. Verify that required lap lengths, stagger and offsets are provided. Verify proper installation of approved mechanical connections per the manufacturer's instructions and/or evaluation reports. Ensure that all welds of reinforcing steel and other weldments are as specified and have been inspected and approved by an approved welding inspector.

4. Formwork, Joints and Embeds

Verify that formwork will provide concrete elements of the specified size and shape. Verify that the location and preparation of construction joints are in accordance with the approved plans, specifications and building code requirements. Verify that the type, quantity, size, spacing and location of embedded items are as specified.

5. Concrete Placement, Protection and Curing

Verify acceptable condition of the place of deposit before the concrete is placed. Verify that methods of conveying and depositing concrete avoid contamination and segregation of the mix. Verify that concrete is being properly consolidated during placement. Verify that concrete is protected from temperature extremes, and determine that proper curing is initiated.

B. Prestressed Concrete Special Inspector

Note – Certification as a Reinforced Concrete Special Inspector is required before certification as a Prestressed Concrete Special Inspector. All items listed above under Reinforced Concrete are considered prerequisite to the knowledge for special inspection of prestressed concrete.

1. General Requirements

Comply with special inspection requirements of the enforcing jurisdiction. Review approved plans and specifications for project details that pertain to special inspection requirements. Notify the contractor of deviations from approved plans and specifications. If the deviations are uncorrected, notify the architect or engineer of record and the building official. Submit progress reports to the architect or engineer of record and the building official, describing tests that were performed and compliance of work. Submit final summary report stating whether work requiring special inspection was in conformance with the approved construction documents and applicable provisions of the building code.

2. Concrete Quality

Verify that individual batch tickets indicate delivery of the approved mix as specified. Verify time limits of mixing, total water added, and proper consistency and workability for placement. Determine the required type, quantity and frequency of tests to be performed on fresh and hardened concrete. Observe sampling of concrete, field testing of fresh concrete and making of test specimens. Provide or arrange for proper specimen identification, site storage and protection, and transportation to the testing laboratory. Provide or arrange for communication of field testing results to the architect or engineer of record and to the building official.

3. Reinforcement

Verify that reinforcing steel and tendons are of the type, grade and size specified and are in conformance with acceptable quality standards. Verify that the reinforcing steel and tendon system are fabricated in conformance with acceptable quality standards. Verify that the condition of tendons at the time of concrete placement are free of oil, dirt and excessive rust, and are properly coated and/or sheathed as specified. Verify that reinforcing steel and tendons are located within specified tolerances, and are adequately supported and secured to prevent displacement during concrete placement. Verify that minimum concrete cover is provided. Verify that placement of reinforcing steel and tendons (or ducts) comply with spacing, profile and quantity requirements, as indicated by the installation drawings and approved plans. Verify that hooks, bends, ties, stirrups and supplemental reinforcement are fabricated and placed as specified. Verify that required lap lengths, stagger and offsets are provided. Verify proper installation of approved mechanical connections per the manufacturer's instructions and/or evaluation reports. Verify that welds have been inspected and approved as specified. Verify that prestressed rock and soil anchors are fabricated and installed as specified.

4. Prestressing and Grouting

Verify that the required concrete strength has been attained prior to transferring prestressing forces. Verify proper equipment calibration. Verify that proper stressing (or tensioning) sequences are used, proper jacking forces are applied, and acceptable elongations are attained and recorded. Verify that tendons and anchorages are properly sealed or otherwise protected as specified. Verify that ducts including inlets and outlets are of the required size, are mortar-tight and are located correctly. Verify that proper grout materials, strength and grouting pressures are used as specified.

5. Formwork, Joints and Embedments

Verify that formwork will provide concrete elements of the specified size and shape. Verify that the location and preparation of construction joints are in accordance with the approved plans, specifications, and applicable codes and standards. Verify that the type, quantity, size, spacing, condition and location of embedded items are as specified.

6. Concrete Placement, Protection and Curing

Verify acceptable condition of the place of deposit before the concrete is placed. Verify that methods of conveying and depositing concrete avoid contamination, segregation of the mix, and displacement of reinforcement, embedments and forms. Verify that concrete is being properly consolidated during placement. Verify that concrete is protected from ambient temperature extremes during placement and curing. Verify that concrete is being cured as specified by approved plans, specifications and applicable codes.

C. Structural Masonry Special Inspector

1. General Requirements

Review approved plans, specifications and submittals for special inspection requirements. Comply with special inspection requirements of the enforcing jurisdiction in accordance with the applicable code, approved plans and specifications. Notify the contractor of deviations from approved plans and specifications. If the deviations are uncorrected, notify the architect or engineer of record and the building official of deviations. Submit progress reports to the architect or engineer of record and the building official, describing tests that were performed and compliance of work. Submit final summary report stating whether work requiring special inspection was in conformance with the approved plans and applicable provisions of the building code.

2. Materials

Verify that brick, block, cement, lime, aggregates, reinforcement, connectors, water, admixtures and other materials are the type specified and approved. Verify that materials are properly stored. Verify that mix proportions, material handling and mixing are in accordance with code requirements. Verify that grout is batched in accordance with approved mix. Determine the required material strengths, type and frequency of tests to be performed. Observe sampling, field testing and fabrication of test specimens. Verify that masonry strength meet approved specifications. Verify proper sample identification, site storage, protection and transportation to the testing laboratory.

3. Masonry Placement

Verify that the condition of substrate is acceptable for placement, that mortar is properly placed and that the masonry units are placed in accordance with the approved plans. Verify that the type, quantity, size, spacing and location of embedded items are as specified. Verify that the location and preparation of movement joints are in accordance with the approved plans, specifications and building code requirements. Verify that the masonry is protected from temperature extremes and adverse weather conditions.

4. Reinforcement and Connector Placement

Verify that the reinforcing steel and connectors comply with required size, spacing, profile, condition and quantity requirements, as indicated by both the approved plans and installation drawings. Verify that reinforcing steel and connectors are placed in the proper location within acceptable tolerances. Verify minimum coverage and clearance to masonry surfaces. Verify that hooks, bends, ties, stirrups and supplemental reinforcement are fabricated and placed as specified. Verify that required lap lengths, stagger and offsets are provided. Verify installation of approved mechanical connections per manufacturers instructions and/or evaluation reports.

5. **Grout Placement**

Verify that grout spaces are free of obstructions and that cleanouts are provided as required. Verify that methods of conveying and placing grout avoid contamination and segregation and comply with time limits and grout lift requirements. Verify that grout is being properly consolidated and reconsolidated during placement.

D. Structural Steel and Bolting Special Inspector

1. **General Requirements**

Review approved plans and specifications for special inspection requirements. Comply with special inspection requirements of the enforcing jurisdiction. Notify the contractor of deviations from approved plans and specifications. If the deviations are uncorrected, notify the architect or engineer of record and the building official of deviations. Submit progress reports to the architect or engineer of record and the building official, describing tests that were performed and compliance of work. Submit final summary report stating whether work requiring special inspection was in conformance with the approved plans and applicable provisions of the building code.

2. **Material Sampling, Testing and Verification**

Verify that the steel shapes and bolts are of the type, size, grade and condition specified on the approved plans and specifications. Verify the required type, quantity, location and frequency of tests to be performed, and witness preparation of properly identified test material samples on all materials. Provide or arrange for documentation and transportation of samples to the laboratory. Verify that required testing is performed on materials as required by applicable standards and specifications.

3. **High-Strength Bolting**

Verify correct type, size and location of bolts and bolt holes, nuts and washers for type of connection specified on approved plans and specifications. Verify protected storage of bolts, nuts and washers as required by applicable standards and specifications. Verify that faying surfaces at connections utilizing high-strength bolts are in compliance with applicable standards. Observe or conduct bolt tension verification tests on required high-strength bolt assemblies. Identify and verify joint type and installation of bolt assemblies per approved plans and specifications. Verify use of the approved method and sequence of bolt tightening.

4. **Steel Framing Observation**

Verify that structural steel frame orientation, details and frame member sizes are in accordance with approved plans and specifications. Verify that column base plates are the designed configuration, have correct size hole and proper clearance for grouting. Verify grout placement and sampling. Verify that base plates are securely seated and fastened in accordance with applicable plans and specifications.

E. Structural Welding Special Inspector

Note- The Structural Steel and Bolting Special Inspector Certification is a prerequisite for receiving the certification for structural welding special inspector.

1. General Requirements

Review approved plans and specifications for special inspection requirements. Comply with special inspection requirements of the enforcing jurisdiction. Notify the contractor of deviations from approved plans and specifications. If the deviations are uncorrected, notify the architect or engineer of record and the building official. Submit progress reports to the architect or engineer of record and the building official, describing tests that were performed and compliance of work. Submit final summary report stating whether work requiring special inspection was in conformance with the approved plans and applicable provisions of the building code.

2. Material Sampling, Testing and Verification

Verify that the steel shapes, base metals, filler metals and gases are of the type, size, grade and condition specified on the approved plans, specifications and Welding Procedures Specifications. Verify the required type, quantity, location and frequency of tests to be performed, and witness preparation of properly identified test material samples on all materials. Provide or arrange for documentation and transportation of samples to the laboratory. Verify that required destructive testing is performed on materials as required by applicable standards and specifications. Verify that required nondestructive examinations are performed as required by applicable standards and specifications.

3. Structural, Reinforcing and Sheet Steel Welding

Verify that the welding equipment and process has the capability to produce the specified welds. Insure that welding equipment is calibrated and appropriate for use with the welding process. Verify and/or witness qualification of welders, welding operators and tackers for conformance with American Welding Society standards and specifications. Verify that welders are qualified to perform the specified work. Verify that the proposed welding procedure for structural steel, reinforcing steel and sheet metal is a standard prequalified procedure, or has been properly qualified and approved. Verify that welding processes, sequences and procedures are followed in accordance with approved Welding Procedures Specifications. Review approved plans and specifications for weld types and locations. Verify that filler materials are stored and handled in accordance with manufacturer and project specifications. Verify that base metal to be welded is properly prepared and oriented. Verify that weldments have proper joint geometry and have backing and start/runoff tabs where required. Inspect to ensure that weld and structural steel repairs are performed in accordance with approved procedures. Verify that fabricated elements are within permissible tolerances. Verify that welds have the specified length and effective throat. Verify that the weld profile meets applicable shape, size and quality requirements.

F. Spray-applied Fireproofing

1. General Requirements

Review approved plan and specifications for special inspection requirements. Comply with special inspection requirements of the enforcing jurisdiction. Notify the contractor of deviations from approved plans and specifications. If the deviations are uncorrected, notify the architect or engineer of record and the building official. Submit progress reports to the architect or engineer of record and the building official, describing tests that were performed and compliance of work. Submit final summary report stating whether work requiring special inspection was in conformance with the approved plans and applicable provisions of the building code.

2. Materials, Preparation, Application and Testing

Verify that the proposed materials, the type specified, are properly stored and have been approved by the registered design professional in responsible charge and the building official. Verify that the substrate has been properly prepared and is free of oil, dirt, scale, loose paint or primer and other materials that may prevent adequate adhesion. Identify the members to be fireproofed and the minimum required coverage and thickness. Verify the condition of the finished application. Determine the required type and frequency of tests to be performed. Observe the sampling, field testing and fabrication of test specimens. Verification that materials are of type specified, properly stored and approved; verification that the substrate has been properly prepared and free of conditions which may prevent adhesion; identification of members to be fireproofed, the minimum required coverage and thickness of the fireproofing, and the condition of the finished application; and determination of the required tests and observation of sampling, field testing and fabrication of test specimens.

(This page is intentionally left blank.)

APPENDIX C

Examples of Special Inspector Qualification Standards

Per International Accreditation Services

AC 291

(This page is intentionally left blank.)

Special Inspector Qualification Standards

The minimum qualifications listed below are from the IAS Document AC291 - Accreditation Criteria for IBC® Special Inspection Agencies and are given for examples of qualifications. Experience is hard to replace with education and where the responsible professional for an agency provides his signature as evidence of competency for the special inspector, it should be respected. Ultimately, the IBC places the responsibility for approval of special inspectors and special inspection agencies upon the building official. In many jurisdictions, the obtaining of the certificate is sufficient for proof of competency.

Experience

1. In order for experience to count toward qualifications, it must be based on verifiable work directly related to the category or type of inspection involved.
2. An engineering degree (BS) plus appropriate in-house training may be substituted for not more than one year of experience. An engineering technology degree (AA) plus appropriate in-house training may be substituted for not more than six months of experience. (Degree experience may not be substituted for more than half of the experience requirements in any category.)
3. Five of more years experience as a qualified special inspector in one or more categories of work may fulfill up to half of the experience requirements in any category, at the discretion of the (agency's) responsible professional engineer.

Certification

Certification, when specified, is intended to mean successful completion of an ICC examination appropriate to the category of work involved.

Special Inspector in Training

1. The intent of this provision is to provide practical opportunities for an inspector to gain the needed experience to qualify as a special inspector.
2. An inspector who does not meet the qualifications for special inspector may be allowed to perform "Special Inspection" at the discretion of the agency's responsible professional engineer, provided one or more of the following conditions are met:
   (a) Individual is working under direct and continuous supervision of a special inspector fully qualified for the type of work involved.
   (b) Individual is working under indirect or periodic supervision of a special inspector, and the scope of work is minor and/or routine and within the capabilities of the individual.
   (c) Individual is specifically approved by the building official.

Section 6.0 Minimum Qualifications for Special Inspectors, AC291

6.1 CONCRETE CONSTRUCTION (PRESTRESSED AND REINFORCED)

PRESTRESSED

6.1.1 Current certification in ICC prestressed concrete inspection and one year of experience; or

6.1.2 P.E. and a minimum one year of direct experience in prestressed concrete construction. Inspector must be qualified under Section 6.1.1 within 12 months of accreditation; or

6.1.3 Bachelor's degree in Civil or Structural Engineering from an accredited institution and a minimum one year of experience. Inspector must be qualified under Section 6.1.1 within 12 months of accreditation; or

6.1.4 ACI Concrete Construction Special Inspector and a minimum two years of experience. Inspector must be qualified under Section 6.1.1 within 12 months of accreditation.

REINFORCED

6.1.5 Current certification in reinforced concrete special inspection by ICC (see note below) and one year of experience; or

6.1.6 P.E. and a minimum one year of direct experience in reinforced concrete construction. Inspector must be qualified under Section 6.1.5 within 12 months of accreditation; or

6.1.7 Bachelor's degree in Civil or Structural Engineering from an accredited institution and a minimum two years of experience. Inspector must be qualified under Section 6.1.5 within 12 months of accreditation; or

6.1.8 ACI Concrete Construction Special Inspector and a minimum one year of experience. Inspector must be qualified under Section 6.1.5 within 12 months of accreditation.

Note: Passing the ICC exam on reinforced concrete special inspection or having the reinforced concrete associate certification will not be considered without meeting the education/work experience requirements by ACI & ICC.

6.2 NONDESTRUCTIVE TESTING (NDT)

6.2.1 Personnel qualified in accordance with nationally-recognized NDT personnel qualifications practice or standard, such as ANSI/ASNT-CP-189 or SNT-TC-1A; or

6.2.2 American Society for Nondestructive Testing (ASNT) Level II and a minimum one year of direct testing experience as determined and approved by an in-house ASNT Level III.

6.3 PIER AND PILE FOUNDATIONS

6.3.1 Current ICC certification in reinforced concrete special inspection in addition to one of the following:

6.3.2 P.E. and a minimum of one year of experience.

6.3.3 NICET III or IV (geotechnical/construction or construction material testing/soils), and a minimum of five years of experience.

6.3.4 NICET CT Certified Engineering Technologist and a minimum of five years of experience.

6.4 POST-INSTALLED STRUCTURAL ANCHORS IN CONCRETE

6.4.1 Current ICC certification in reinforced concrete special inspection.

6.4.2 Current ICC certification as a residential or commercial building inspector, as applicable, and a minimum two years of experience related to the activity being inspected; or

6.4.3 P.E. and a minimum one year of experience related to the activity being inspected. Inspector must be qualified under Section 6.4.1 within 12 months of accreditation; or

6.4.4 Bachelor's degree in Civil or Structural Engineering from an accredited institution and a minimum two years of experience related to the activity being inspected. Inspector must be qualified under Section 6.4.1 within 12 months of accreditation.

6.5 SOILS

6.5.1 NICET II, III, IV or CT (geotechnical/construction or construction material testing/soils) and a minimum two years of experience; or

6.5.2 Technician with a minimum three years of documented experience directly related to soils testing and inspection under a licensed P.E. Inspector must be qualified under Section 6.5.1 within 12 months of accreditation; or

6.5.3 Bachelor's degree in Civil or Structural Engineering/Geo-tech/Geologist from an accredited institution and a minimum one year of experience. Inspector must be qualified under Section 6.5.1 within 12 months of accreditation; or

6.5.4 P.E. in civil engineering or equivalent P.E. and a minimum one year of experience. Inspector must be qualified under Section 6.5.1 within 12 months of accreditation.

6.6 SPRAY-APPLIED FIRE-RESISTANT MATERIALS

6.6.1 Current ICC certification as a spray-applied fireproofing special inspector and a minimum of one year experience; or

6.6.2 P.E. and a minimum one year of experience in fireproofing applications. Inspector must be qualified under Section 6.6.1 within 12 months of accreditation; or

6.6.3 Bachelor's degree in Civil or Structural Engineering from an accredited institution and a minimum two years of experience in fireproofing applications. Inspector must be qualified under Section 6.6.1 within 12 months of accreditation.

6.7 STEEL (HIGH-STRENGTH BOLTING AND WELDING) BOLTING

6.7.1 Current ICC certification as a Structural Steel and Bolting Special Inspector and a minimum one year of experience.

Note: Current certifications for Structural Steel and Welding Special Inspector are valid until the date of expiration.

6.7.2 AWS Certified Welding Inspector (CWI), and

6.7.3 Current ICC certification as a Structural Steel and Welding Special Inspector and a minimum one year of experience.

6.8 MASONRY CONSTRUCTION

6.8.1 Current ICC certification in masonry and a minimum one year of experience; or

6.8.2 P.E. and a minimum one year of relevant experience. Inspector must be qualified under Section6.8.1 within 12 months of accreditation; or

6.8.3 Bachelor's degree in Civil or Structural Engineering from an accredited institution and a minimum two years experience. Inspector must be qualified under Section 6.8.1 within 12 months of accreditation.

6.9 STRUCTURAL WOOD CONSTRUCTION

Current ICC certification as a commercial or residential building inspector, as applicable, and

6.9.1 A minimum two years of direct experience in engineered wood products; or

6.9.2 A minimum five years of direct experience as a journeyman carpenter.

6.10 EXTERIOR INSULATION AND FINISH SYSTEMS (EIFS)

6.10.1 Current ICC certification as a reinforced concrete special inspector; or

6.10.2 Current ICC certification as a residential or commercial building inspector, and a minimum of two years of experience related to the activity being inspected; or

6.10.3 Bachelor's degree in Civil or Structural Engineering from an accredited institution and a minimum two years of experience related to the activity being inspected. Inspector must be qualified under Section 6.10.1 within 12 months of accreditation; or

6.10.4 NICET CT Certified Engineering Technologist and a minimum five years of experience. Inspector must be qualified under Section 6.10.1 within 12 months of accreditation.

6.11 SPECIAL CASES

6.11.1 Current ICC certification as a special inspector and a minimum two years of experience related to the activity being inspected; or

6.11.2 Bachelor's degree in Civil or Structural Engineering from an accredited institution and a minimum two years of experience related to the activity being inspected. Inspector must be qualified under Section 6.11.1 within 12 months of accreditation; or

6.11.3 P.E. and a minimum one year of experience related to the activity being inspected. Inspector must be qualified under Section 6.11.1 within 12 months of accreditation.

Exception: Individuals who have proven expertise in a field of specialty, either through education or field experiences of not less than five years, may be considered as meeting criteria to conduct one or more classes of specialty inspections.

Reference Abbreviations and Recognized Certifying Agencies

1.	AA	Associate of Arts
2.	ACI	American Concrete Institute
3.	ACIA	American Construction Inspectors Association
4.	ASTM	American Society for Testing and Materials
5.	ASNT	American Society for Nondestructive Testing
6.	AWS/CAWI	American Welding Society/Certified Associate Welding Inspector
7.	AWS/CWI	American Welding Society/Certified Welding Inspector
8.	BS	Bachelor of Science
9.	IBC	International Building Code
10.	NICET	National Institute for Certification of Engineering Technologists
11.	NRCA	National Roofing Contractors Association
12.	IAS	International Accreditation Service
13.	PDS	Professional Development Services (Division of ICC)